Illustration of the *Galileo* probe on its way to Jupiter

Jupiter

Steve Potts

A⁺

Smart Apple Media

COPYRIGHT

☼ Published by Smart Apple Media

1980 Lookout Drive, North Mankato, MN 56003

Designed by Rita Marshall

Printed in the United States of America

☼ Pictures by Photo Researchers (Julian Baum/Science Photo Library, Chris Bjornberg/Science Source, Lynette Cook/Science Photo Library, Library of Congress/Science Source, NASA/Science Source, Science Photo Library, Detlev Van Ravenswaay), Tom Stack & Associates (TSADO/NASA)

☼ Library of Congress Cataloging-in-Publication Data

Potts, Steve. Jupiter / by Steve Potts. p. cm. – (Our solar system)

Includes bibliographical references and index.

☼ ISBN 1-58340-097-4

1. Jupiter (Planet)–Juvenile literature. [1. Jupiter (Planet)] I. Title.

QB661 .P68 2001   523.45–dc21   2001020127

☼ First Edition 9 8 7 6 5 4 3 2 1

Discovering Jupiter

No one was more important to the Romans than their chief god, Jupiter. Perhaps that is why they named the largest planet in our solar system after him. The fifth planet from the Sun, Jupiter is the king of planets. It would take more than 1,300 Earths to equal the size of Jupiter. ☀ Like the planet Venus, the Sun, and Earth's moon, Jupiter is a very bright object in the sky. Italian **astronomer** Galileo Galilei used his primitive telescope to peer at Jupiter. In 1610, he dis-

Astronomer and mathematician Galileo Galilei

covered its four largest moons: Io, Europa, Ganymede, and Callisto. These are now known as the Galilean moons. ⚙ When Galileo found these four moons, it was a major discovery.

Many people at the time believed that Earth was the center of the solar system. Instead, Galileo discovered that the Sun was the center of our solar system. By finding Jupiter's moons, Galileo proved that Earth was not the only planet that had a moon orbiting it.

One day on Jupiter is 9.8 Earth hours long. One year is 11.86 Earth years long.

Jupiter's Characteristics

Astronomers have found that Jupiter is swept by fast winds and great storms. For more than 300 years, people have

Jupiter and its four large Galilean moons

Jupiter's Great Red Spot is a huge, violent storm

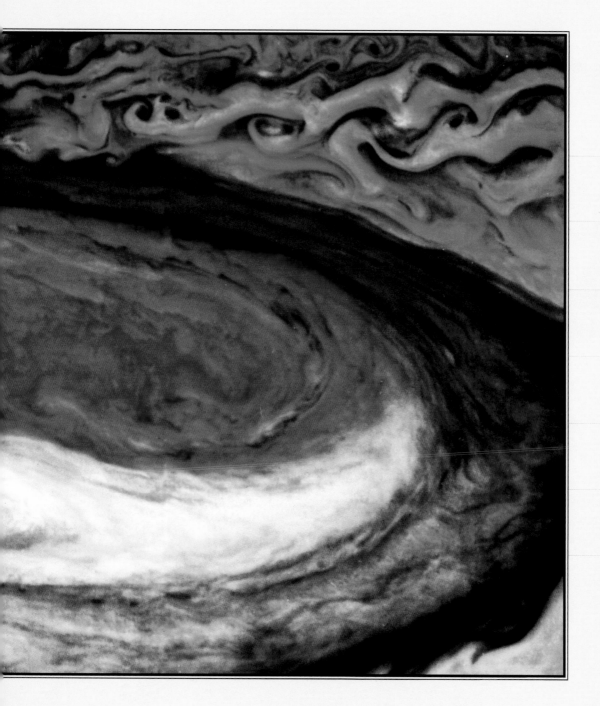

been able to see what astronomers call the Great Red Spot

through telescopes. This "spot" is actually a huge storm the

size of two Earths. It looks like a big, red, egg-shaped cloud.

The storm contains winds of up to 340 miles **Jupiter's weight is 2.5 times greater than all the other planets combined.** (545 km) per hour that blow along Jupiter's equator. ☀ In addition to its huge storms and high winds, Jupiter differs from Earth in

another way. It is made up of very different material from

Earth. Jupiter, like the other outer planets (Saturn, Neptune,

Uranus, and Pluto) is made mostly of gases and liquids.

Scientists think that Jupiter has a small center made of rock. All the rest of this massive planet is gas and liquid. The tops of Jupiter's gas-and-liquid clouds can be seen through a

Scientists think Jupiter has a rocky core surrounded by gases

telescope. ☀ Some scientists say that there are probably inner circles of metallic hydrogen and liquid hydrogen surrounding the planet's center. Metallic hydrogen is a form of hydrogen that can carry electricity. An 800-mile-thick (1,280 km) layer of gases, in the form of a cloud, surrounds these inside layers. ☀ Astronomers believe that this mixture of hydrogen and helium was the same gas that surrounded all the planets when they were originally formed. This mixture also creates a very cold **environment**. Temperatures at the top of this cloud

Probes have detected oxygen and water on Jupiter's moon Ganymede.

mixture are –230 °F (–145 °C). ☀ These thick layers of

clouds and gases change as winds blow them in different

directions above the planet's surface.

The thick, cold clouds of gases around Jupiter

Moons and Rings

One of the most fascinating facts about Jupiter is that it has 16 moons. They all orbit the planet at different distances from its surface. Besides the four Galilean moons, modern space explorers have found 12 more. ✷ All of the moons differ from each other. Some are made of iron and rock. Others are mostly ice. Scientists believe that one of the moons, Europa, may have an active ocean beneath its icy

The *Galileo* space probe traveled six years and half a billion miles to reach Jupiter.

Some of Jupiter's moons are the size of planets

surface. ☀ Astronomers recently learned that Jupiter has a

ring system similar to that of Saturn. Though much smaller, it

is very bright because it is made up of fine dust that reflects

light. The main ring is about 4,350 miles (7,000 km) wide.

Exploring Jupiter

In recent years, astronomers have discovered more about Jupiter by sending space probes to the distant planet. Probes are small spacecraft that use computers and cameras to study objects in space. *Pioneer 10* was the first probe to reach Jupiter in 1973. Later, *Voyager 1, Voyager 2,* and *Ulysses* visited the planet. ✳ In 1989, the probe *Galileo* arrived at Jupiter and began recording information. A mini-probe was dropped from it in 1995 and sent toward Jupiter's surface. For 58 minutes it

The *Galileo* probe gathered amazing images of Jupiter

took extremely detailed pictures that were sent back to Earth.

These pictures gave scientists a close look at objects as small

as 40 feet (12 m) across. Eight hours later, it burned up in

Jupiter's atmosphere, which was recorded to A belt above Jupiter's

be 3,400 °F (1,870 °C). ☀ Scientists will clouds contains the strongest

continue to study Jupiter using probes, with radio waves in the solar

the hope of learning much more about this system.

amazing planet.

Jupiter is surrounded by a small ring of fine dust

Jupiter (lower left) is by far the largest of the planets

23

INFORMATION

Index

Words to Know

astronomer—a scientist who studies planets, asteroids, moons, and other objects in the solar system and space

environment—the living and non-living surroundings of an organism or a group of organisms

equator—the imaginary ring around the center of a planet that divides the northern half from the southern half

Read More

Bond, Peter. *DK Guide to Space.* New York: DK Publishing, 1999.

Couper, Heather, and Nigel Henbest. *DK Space Encyclopedia.* New York: DK Publishing, 1999.

Furniss, Tim. *Atlas of Space Exploration.* Milwaukee, Wisc.: Gareth Stevens Publishing, 2000.

Internet Sites

Astronomy.com
http://www.astronomy.com/home.asp

NASA: Just for Kids
http://www.nasa.gov/kids.html

The Nine Planets
http://seds.lpl.arizona.edu/nineplanets/nineplanets

Windows to the Universe
http://windows.engin.umich.edu